GREAT SCIENTISTS

ARISTOTLE

Belitha Press

This edition published in 2003 by
Belitha Press
A member of **Chrysalis** Books plc
64 Brewery Road, London N7 9NT

Typeset by Chambers Wallace, London
Printed in Malaysia

British Library Cataloguing in Publication Data
for this book is available from the British Library.

ISBN 1 84138 640 5

Cover montage images provided by Mansell
Collection, Mary Evans Picture Library and
Image Select/Ann Ronan Picture Library

Illustrations: Tony Smith 10/11, 16/17, 21;
Rodney Shackell 15, 16 top left, 18, 19, 22, 23
Editor: Rachel Cooke
Designer: Andrew Oliver
Picture researcher: Juliet Duff
Consultant: Dr Catherine Osborne

Acknowledgements

Photographic credits:
Ancient Art & Architecture Collection 9 top right
    and bottom, 27 top
Bridgeman Art Library 4 bottom (British Museum,
    London), 5 bottom (British Library, London),
    6 top (National Museum, Athens), 12/13
    bottom (Vatican Museums and Galleries,
    Rome), 15 bottom right (Österreichische
    Nationalbibliothek, Vienna), 20 bottom
    (Musée des Beaux-Arts, Lille, Giraudon), 26
    (Lauros/Giraudon/Bibliotheque Municipale,
    Rome)
C. M. Dixon 14 top
E. T. Archive 12 top, 13 top (Capitoline Museum,
    Rome), 27 bottom left (Royal College of
    Surgeons)
Mary Evans Picture Library 6 bottom, 8 all,
    20 top, 24
Michael Holford 4 top, 10 top
Image Select/Ann Ronan Picture Library 7
Mansell Collection 9 top left, 25
Andrew Oliver 5 top
Oxford Scientific Films 18 bottom (David
    Thompson), 19 top (Scott Camazine), 19
    centre left (Kim Westerskov)
Planet Earth Pictures 17 top, 19 centre right,
    19 bottom left
Popperfoto 22 bottom
Science Photo Library 27 bottom right (Peter
    Menzel)
John Walmsley Photo Library 23 top

Contents

The philosophy of Ancient Greece is not its only legacy to us today. Some of its buildings and works of art still survive, 2,500 years later. The Temple of Apollo (above) at Delphi and the amphora or vase (below) are just two examples of this rich heritage.

Introduction

Science today is a logical, step-by-step process. It involves making a proposal or **hypothesis**, carrying out tests and experiments, measuring and recording the results, and seeing how these fit the hypothesis. This leads to knowledge based on factual evidence. Science is also practical. Its application has produced countless inventions and machines, from levers, wheels and clocks, to computers, cars and television.

However, when science was first considered in ancient times, it was not for its uses in daily life. For the ancient Greeks, in particular, it formed part of their study of **philosophy** – the ways we think, the nature of knowledge and existence, and the search for truth.

Foremost among those ancient Greek philosophers was Aristotle, who lived more than 2,300 years ago. He thought and wrote about a huge range of subjects, and in particular his philosophy of science, from nature study to physics, was to shape scientific thought for centuries to come. Many of his ideas are long out of date. But, if a person had to be singled out, perhaps as the "Grandfather of Science", one of the favourites would be Aristotle.

Chapter One
Aristotle's Inheritance

Before looking at the life and works of Aristotle, it helps to understand something of the world in which he lived, and how information about him, and his times, has reached us today.

Much of our knowledge about the ancient Greeks comes from their books and writings. Few original texts have survived. Many have been copied, translated, altered, copied again, and so on, each version reflecting the thoughts and ideas of the people making it. The same applies to information that has been handed down in spoken form. It is like a game of Chinese whispers.

What we think we know about those far-off times may not be exactly as it was. This results in disagreements between the modern versions of who said what, and when. Sorting through the different sources of information is part of the ongoing search for knowledge and truth.

Early scientific knowledge was applied to construct the Great Pyramids of Ancient Egypt.

Science before the Greeks

In the times of Ancient Mesopotamia and Egypt, more than 3,000 years ago, science appeared in a practical and useful form. People wanted to construct great buildings, such as the Pyramids, and find routes across deserts and seas. So they devised ways to plan and measure, to work with numbers in arithmetic and shapes in geometry, to know about materials and engineering, and to study the starry skies in astronomy.

This illustration of Aristotle comes from a 13th century Arabic work, The Description of Animals *by Ibn Bakhtishu. Some of Aristotle's work only survived through the translations and interpretations of Arabic scientists and philosophers.*

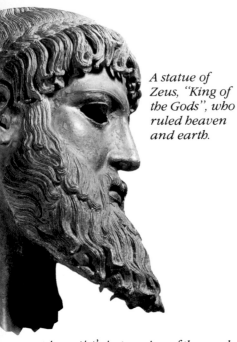

A statue of Zeus, "King of the Gods", who ruled heaven and earth.

Ancient Greece

The ancient Greek civilization began to emerge about 1400 BC. The Greeks were seafaring traders. They established cities, and produced beautiful buildings and fine works of art. They developed complex systems of laws, politics and government.

Some Greeks could speak freely and vote for their government representatives. But not all. There were many peasants and slaves, who had few personal rights. Cheap slave labour meant that there was little need to improve on the innovations made in Egypt and Mesopotamia (see panel on page 5). The sciences of engineering and technology made slow progress.

Greek religions and traditions involved numerous gods and spirits. Zeus was their chief. He lived on Mount Olympus and threw lightning bolts in anger. Gods, rather than science, helped to explain events in the natural world. The Sun and Moon, plants and animals, storms and lightning, diseases and plagues, were their doing.

An artist's impression of the market place in Athens, the principal city of Ancient Greece. The city was dominated by the walled hill of the Acropolis, enclosing the Parthenon, temple to Athena, goddess of wisdom.

The Greek Philosophers

The Greeks visited many lands, and encountered different religions and customs. Some began to question their own beliefs and assumptions, and this gave rise to the study of philosophy.

The term "philosophy" comes from an ancient word *philosophos,* which is Greek (of course) for "lover of wisdom". Philosophy is a complex subject. It tackles questions that most people rarely consider. How do we know what we know? Where do our thoughts, beliefs and knowledge come from? How do we know or decide what is good or bad, right or wrong, true or false? What do words such as knowledge, reality and truth actually mean?

Philosophy began in Ancient Greece around 600 BC. The first philosophers looked for general underlying principles, that would explain the world around them. They introduced methods of reasoning, **logic**, observation, and even some experiments. Into these surroundings came two of the greatest philosophers – Plato, and his pupil Aristotle.

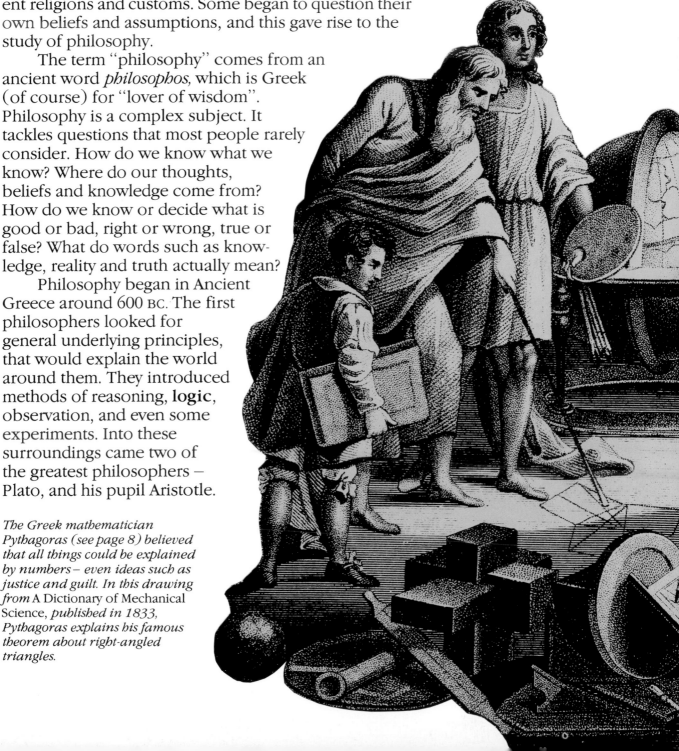

The Greek mathematician Pythagoras (see page 8) believed that all things could be explained by numbers – even ideas such as justice and guilt. In this drawing from A Dictionary of Mechanical Science, *published in 1833, Pythagoras explains his famous theorem about right-angled triangles.*

Great Greeks

Many notable philosophers lived before Aristotle. Their thoughts and ideas seem strange today, even laughable. But they were great thinkers of their time, and they helped to shape Aristotle's work.

- Democritus (460-370 BC, below) suggested that all matter was made of tiny units that were indivisible, now known as **atoms**.

- Anaxagoras (500-429 BC, above) believed that every substance was a mixture containing a little of everything, and that you could never obtain a pure **element**.

- Hippocrates of Cos (460-377 BC, below) is often called the "Father of Medicine" (see opposite).

- Pythagoras of Samos (540-480 BC, above) held that mathematics could answer most questions, and that "all things are numbers".

- Thales of Miletus (636-546 BC, below) was the first known scientific philosopher. He believed that all things came from water, and consist of water in different forms.

- Anaximander (610-545 BC), pupil of Thales, wrote one of the first scientific books, about the history of the Universe.

- Anaximenes (570-500 BC) said that everything was made of different forms of air, and that rainbows were natural and not godly.

Chapter Two
The Early Years

Aristotle was born at Stagira in Thrace in the year 384 BC. His father Nicomachus was physician (doctor) to King Amyntas of Macedonia, a region of northern Greece. Nicomachus belonged to a large family of physicians and healers who could supposedly trace their ancestors back to the famous Hippocrates of Cos.

Aristotle grew up in the court of King Amyntas. He was a friend of the King's son, Philip (left). This boy prince later succeeded his father and became Philip II of Macedonia.

A Keen Eye

As a boy, Aristotle probably watched his father treating patients and making medicines from parts of plants and animals. From this, Aristotle may have gained his interests in nature study, biology, and **anatomy** – which concerns the internal structures of living things. During this time, too, he may have developed his keen powers of observation, which were so important in his later studies of the natural world.

A Greek carving from the 4th century BC of a physician treating his patient.

Ancient Greek surgical knives and blood-catching cups.

Greek Medicine

The ancient Greeks, most importantly Hippocrates and his colleagues, began the era of scientific medicine. Before their time, and in many places since, medicine was tied up with religion and superstition. The physician rarely even examined the patient. He guessed the illness from a religious "sign".

The ideas of Hippocrates and his followers are an early example of scientific methods applied to the real world. They taught that the physician should examine and observe the patient, and work out the illness from the symptoms. He should give medicines only when necessary, and check the patient later, to see the effects of the medicine and the course of the disease. Only by observing, and keeping records, could physicians gather real knowledge and make progress in medicine.

This Greek vase, 2,400 years old, shows a boy being taught to ride a horse. Many young men practised physical skills such as riding, wrestling and throwing.

Growing up in Ancient Greece

For Greek families such as Aristotle's, life was fairly comfortable. They had money and servants. The boys learned subjects such as mathematics, politics, philosophy, logic and **ethics** (see page 22). They learned about the ways of the gods. There was no science as we recognize it today. As they became older, they would worship at the temples, and attend the theatre to watch plays, musical concerts and dancing.

Aristotle's parents died when he was a child. So there may have been less encouragement – or pressure – for him to become a physician, in the family tradition. Instead his education took a different direction. For in 67 BC, when Aristotle was 17 years old, he travelled to Athens, Greece's principal city. Here he joined the academy (school) of the renowned philosopher Plato.

Tales of Aristotle's teenage years say that he wasted his family's money on high living, and he was easily provoked to an argument. However, by the age of 17, he seemed keen to continue his education at the highest level. So he travelled to the centre of the Greek world, Athens, to study with the famous Plato.

11

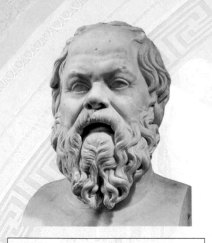

Chapter Three
The Academy in Athens

Plato had founded the Academy twenty years before Aristotle arrived there. It was a type of "school" where people could think, develop understanding, and acquire knowledge. It was sited in a grove on the outskirts of Athens. This is where Aristotle would have spent much of his day.

A public gymnasium seems to have been the base for the discussions. The site was dedicated to the memory of Herademos (Akademos), a hero of the Trojan Wars. This probably gave rise to the name Academy, and to our word "academic". This means to do with learning and knowledge, or to be concerned with theories and ideas rather than practical topics.

Socrates (469-399 BC)

The Traditions of Socrates

The philosopher Socrates sought after true knowledge and wisdom. He lived all his life in Athens, and wrote nothing himself. We know of him mainly through his pupil Plato.

Socrates always asked questions and searched for definitions. He believed that people, asked in the correct way, could find the answers within themselves, and so teach themselves. From this came the Socratic tradition of furthering knowledge by asking questions, that is, **dialogue**.

Socrates' personal life is shrouded in mystery. He was supposedly unattractive and behaved oddly, and earned a small wage as a stonecarver. At the age of 70, he was accused of misleading and corrupting young people in Athens, by his questioning. He was sentenced to death by drinking poison.

This huge wall-painting by the Italian artist Raphael, completed in 1511, is called School of Athens. *It depicts teachers and students from the time of Plato and Aristotle, the two figures at the centre of the picture.*

Mind over Matter

The Academy was not like the typical schools we have today. There were few set subjects and lessons. Little importance was put on practical studies such as engineering, physics and chemistry. There were few attempts to carry out observations or experiments, and investigate the real world of objects and events.

Instead, Plato encouraged his students to let their minds and thoughts roam free. They strove to break away from daily realities, from what happens in the physical world, and from the things we see, hear and feel around us. Their common aim was the pursuit of knowledge and understanding, and they followed their purpose almost like a religion.

The Academy also trained young men for politics and leadership, and advised rulers. It continued long after the time of Plato and Aristotle, until AD 529, when it was closed by the Byzantine emperor, Justinian I.

Plato (427-348 BC)

Platonic Philosophy

After the death of his teacher Socrates, Plato travelled until 387 BC, when he returned to Athens and founded the Academy.

Plato's writings are mainly in the form of dialogues, presenting fictional conversations between characters. Socrates was usually the main speaker. Plato's dialogues cover a wide range of topics, from justice and politics, to the origin and nature of the universe.

Plato talked about things he called "Forms". They explain our ideas about goodness, justice, beauty and so on, and how we know the truth about such things. From his work came the Platonic tradition, in which matters of the mind, and the ideas behind words, were the true basis of existence. The physical world, which most people today think of as "real", was less important.

Another example of the lasting influence of the Academy, this mosaic was unearthed from the Roman town of Pompeii and shows Plato and his pupils in deep discussion. Pompeii was buried under layers of ash from the sudden eruption of the volcano Mount Vesuvius in AD 79.

Aristotle at the Academy

Aristotle became a pupil at the Academy, then a teacher. He stayed for about 20 years. During this time, Plato taught him and became his friend. Aristotle was inspired by Plato, and learned all he could from him.

Much of the discussion at the Academy was in the tradition of a dialogue, or discussion between two people. This was pioneered by the philosopher Socrates (see page 12). One person asks a question. The other talks about it, suggests answers, and asks more questions, and so on. Gradually the answers should emerge.

Debates in the form of a dialogue could be very thought-provoking. Plato often asked if things we see, smell and touch are truly real. What do we mean by "real", when the only evidence we have is our own senses? Could things exist only in our minds, in our imaginations? The need to find answers to these questions may have stimulated Aristotle in his later studies and observations of animals and plants.

While at the Academy, Aristotle wrote several works, in the form of dialogues, but only fragments remain.

Leaving the Academy

As Aristotle's own knowledge increased, he began to question Plato's views and the methods of the Academy. Matters came to a head when, in 347 BC, Plato died. His nephew Speusippos took over the Academy. He was a mathematician in the tradition of Pythagoras (see page 8), who believed that the world and everything in it could be explained by numbers and sums. Aristotle had become restless, and he could not agree with Speusippos and his views. So he set off on his travels.

The Four Elements

Aristotle developed a great interest in why things move as they do. Why do stars arc across the night sky? In *On the heavens*, he discussed one of the common beliefs of his time. All matter was made of what we might today call "elements". There were four elements: earth, air, fire and water.

Aristotle suggested that each element had its own type of movement, in a certain direction. But stars and heavenly bodies seem to have a circular motion, different from the movements of the four elements. So he proposed that they were composed of a fifth element, **aither** (or ether), that naturally moved in a circle.

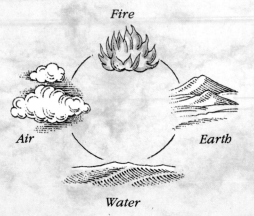

The four elements as illustrated in a late medieval book. Aer is air, Ignis is fire, Tra is earth, and Aqva is water.

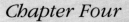

Chapter Four

The First Naturalist

Aristotle was away from Athens for 12 years. First he went to Assos, to the court of Hermias of Atarneua, part of what is now Turkey. Here he studied politics and biology, taught pupils and advised the rulers. He married the niece of Hermias.

He established academies in Assos and also in Mitylene, where he moved to after three years. He also travelled to islands such as Lesbos.

Doing, Not Just Thinking

During this time, Aristotle carried out many observations and studies on nature, especially fish and seashore creatures. He cut them open, identified their parts and **organs**, and suggested what these might do.

This way of working was very unusual for the time. Most of his colleagues did not "get their hands dirty". They were thinkers, not doers. They rarely carried out tests, or observed the real world closely, or made records of what they saw. Aristotle was establishing a whole new tradition, which for us today, is a central part of science.

The eventual results were Aristotle's famous works on *The Parts of Animals, The Natural History of Animals* and *The Reproduction of Animals.*

Some of the places visited by Aristotle during his time away from Athens. Modern biologists have been able to guess at parts of his travels from some of the animals he described, which are only found in certain small areas.

Standing the Test of Time

Much of Aristotle's work has been important and admired in the history of science, but it has now become out of date. However, his studies in nature and biology have stood the test of time extremely well.

This is partly because of Aristotle's emphasis on first-hand observation and experience. This was such a change from the traditions of the time, that in *The Parts of Animals*, he wrote several chapters justifying his methods.

In *The Natural History of Animals*, Aristotle made the first serious attempt to group animals according to the features they had in common. He looked for an underlying organization in the natural world.

Aristotle was particularly keen on seashore and sea animals, such as fishes, starfishes and seabirds.

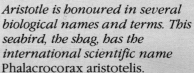

Aristotle is honoured in several biological names and terms. This seabird, the shag, has the international scientific name Phalacrocorax aristotelis.

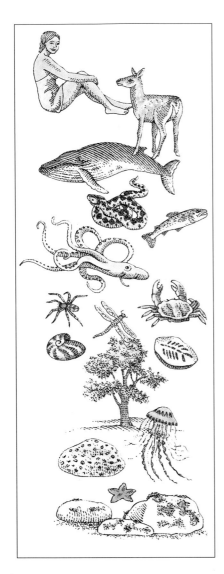

To describe the underlying order of nature, Aristotle devised his "Ladder of Nature". The basic groupings – humans, mammals, whales, reptiles and fishes, octopuses and squids, jointed shellfishes, insects, and so on – are still much the same today.

Red Blood and No Blood

Aristotle's basic classification had two groups, or divisions. One was animals with red blood, such as fish, snakes, birds and mammals. We now call them **vertebrates**, or animals with backbones.

His other group was animals without blood. These are now known as **invertebrates**, or animals without backbones, such as worms, crabs and starfishes. We now know that they have a type of blood, but their blood systems are very different from those of vertebrates.

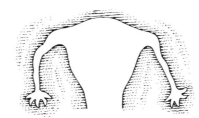

Aristotle's drawing of the uterus (womb) of a female mammal. It is one of the earliest surviving diagrams of anatomy – the study of the structure of body parts.

Anatomy and Embryology

Aristotle's knowledge of anatomy was also first rate. He described the tiny tube that, in mammals, connects the inside of the ear to the back of the throat. This tube was not described again until about 1550, by the Italian Bartolomeo Eustachio. We now call it the Eustachian tube. This is one of many similar examples.

Aristotle put about 500 animals into 8 groups or classes. His studies of fishes, in particular, were careful and detailed. So were his descriptions of **embryology** – how chicks develop inside eggs, and mammal babies inside their mothers. Again, they were hardly improved until the 1600s, almost 2,000 years later.

Aristotle opened birds' eggs to study how a chick first forms.

A Pioneer Biology Book

For its time, *The Natural History of Animals* contains some amazing insights into the natural world.

• Aristotle said correctly that dolphins (above) were air-breathing mammals, and not fishes, as people believed.

• It was thought that the hyaena (below) was an **hermaphrodite**, that is, both male and female in the same animal. Aristotle showed there were male hyaenas and female ones, just like other mammals.

• He saw that in a honeybee hive (above), there was only one queen, though he called it the "king" or "leader". His descriptions of hive life were not bettered until the 1700s.

• He described how a cuttle-fish (above) digested its food.

• He realized that animal parts or organs were suited to doing certain jobs, such as long legs for fast running. He said: "Nature makes the organs to suit the work they have to do". Recognition of this process, adaptation, later led to the modern theory of **evolution**.

Aristotle also made some errors. These are perhaps understandable, given the range of his studies, and the beliefs and traditions of his world. Many can be explained because he had no **microscope**, or he did not travel to the right places.

• He believed that some young animals appeared from mud and water, without parents. He failed to see their eggs which are so tiny, they can only be seen with a microscope – which was not invented for 1,900 years.

• He said that eels never breed. We now know that they do, but in the Sargasso Sea in the West Atlantic, which was far distant from Ancient Greece.

• He thought a goat (above) would be male or female depending on the way the wind blew when its parents mated.

• He stated that the site of intelligence was the heart (above), not the brain.

Alexander and Aristotle

As a boy, Alexander was fascinated by the heroes of the great poet Homer. In 336 BC he took the throne as Alexander III of Macedonia, won control of the army, and went on to conquer much of the known world.

Under Alexander's rule, academies grew up in the tradition of Aristotle and the Greek philosophers in faraway places, such as at Ai Khanoum in present-day Afghanistan. Alexander remembered his teacher, and sent back exotic animals and plants for Aristotle's study. Alexander died at the age of 33, trying to conquer Arabia.

Alexander the Great, on his white horse Bucephalus, leads his armies to victory, in this early 1800s painting by Watteau.

Tutor to the Great

In 356 BC Aristotle's boyhood friend, Philip II of Macedonia, and his wife, Olympias, had a son. He was named Alexander. In 343 BC, Aristotle was asked back to Philip's court, and became tutor to Alexander. He taught the boy mathematics, navigation, astronomy, biology, logic and other subjects.

Some of the Greek philosophers had suggested the idea of training a philosopher-king, a leader to spread their beliefs and make the world a better place. Could Alexander be the one?

The pupil was clever, but he did not take up philosophy or the other subjects. He took up war and conquest. He became known as Alexander the Great. By the time he was 30 years of age, he controlled half of the world known to the Greeks.

Chapter Five
Aristotle's Own School

In 334 BC Aristotle returned from his travels to Athens, and set up his own academy at the Lyceum, a temple of Apollo, the sun god. It was also called the Peripatetic School, from the Greek word *peripatein*, meaning "to pace to and fro". The philosphers taught and discussed while walking around the courtyard.

In the mornings Aristotle would teach his students about topics such as logic and philosophy. In the afternoons he would give public lectures in **rhetoric**, politics and ethics.

A Scheme for Science

Most of Aristotle's writings that have survived date from this period. They were mainly in the form of lecture notes for use in the Lyceum school. They include works on *Physics* and *Metaphysics*. The wide-ranging *Organon* set out his ideas of knowledge and science, and how investigations should be approached and carried out.

He tried to develop the views of people such as Plato, but moved away from pure thought in the mind. He wanted to deal more with the real world and the objects and events in it.

After Plato's death, teachings at the Academy in Athens became limited. Aristotle widened the range and introduced new subjects, such as nature study.

Stands to Reason

Some of Aristotle's great achievements were in the areas of reasoning and logic.

He set out the guidelines for argument and **deduction**. This is the process of thinking and reasoning from a starting point, to reach a conclusion or "answer". We sometimes do this by instinct. We feel that if one thing is true, it follows that something else is also true. But scientists and philosophers must be organized, thorough and consistent in their methods. They cannot simply assume or accept a proposal, without examining it in detail, using reasoning and logic.

The Key is in the Connection

In particular, Aristotle developed the type of reasoning we call **syllogism**. Here is a simple example. All birds have feathers. All parrots are birds. So it follows that all parrots have feathers. This reasoning is of the type: All Bs are Cs. All As are Bs. So all As are Cs.

Note that the order and relationship of information is important. Read the statements about parrots, birds and feathers again. Could you make the deduction that all birds are parrots?

Syllogism does not bring to light any new facts. But it does bring something new: a connection. It is a method of scientific reasoning or confirming, known as formal proof or demonstration.

It's a fact that all parrots are birds. But logical reasoning tells us that we cannot state the reverse. All birds are not necessarily parrots.

Breadth and Depth

Aristotle thought and wrote on such a broad range of subjects, and went into them so deeply, that only a few can be squeezed into a small book on science discoveries. They include:

● Logic, understanding language and the process of reason.
● Physics, understanding the physical and natural world.
● **Metaphysics**, the ideas of existence and "being", their substances and qualities.
● Ethics, dealing with good and bad, morals and politics.
● Rhetoric, involving the power and persuasive use of words in speech, poetry, drama and creative writing. Rhetoric was the counterpart of **dialectic** (opposite).

Some of these words have changed their meanings over the years.

Martin Luther King, 1960s civil rights leader, used rhetoric.

Language and Science

During Aristotle's time, "logic" meant the art and process of using language, to reason from one fact or statement to another. Facts do not simply exist. They must be expressed and understood in terms of words and language. In part of the *Organon*, Aristotle tried to analyze the way language is used as a means or an instrument (*organon*) to express knowledge.

To help in his work, he developed his own versions of **dialectic**. This is a process of critical reasoning. It involves the methods we use, to assess if something is true or valid. These include reason and deduction, and considering arguments both for and against.

Stating the Obvious?

Aristotle laid out the guidelines for dialectic in his *Topics*. He gave credit for beginning dialectic to the philosopher Zeno of Elea, more than a hundred years earlier. To commence an argument, Zeno devised a series of paradoxes. These are statements which contradict what appears to be obvious.

There are many types of logic in use today. They include computer programming logic, for writing programs, and machine logic, the electronic workings inside the computer.

Catching the Tortoise

One paradox involves a race between the Greek hero Achilles and a tortoise. Achilles runs ten times faster than the tortoise, but the tortoise has a start of ten metres. Can Achilles ever catch it?

Achilles runs ten metres. In this time, the tortoise moves one metre. So Achilles runs this metre. But the tortoise has moved a further 10 centimetres. And so on. The distances get small, but does Achilles ever really catch up? Aristotle discusses the paradoxes in his work *Physics*, and they have been fascinating people ever since.

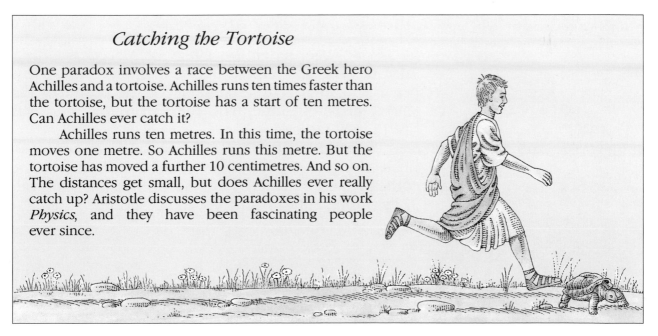

The Four Causes

Aristotle recognized the importance of change, and the processes by which it happens, as well as the results. He said that we cannot truly understand unless we know four different "wherefores" or "reasons why". These have become known as the Four Causes. They can be applied to an everyday object as follows:

● What is it? A kitchen knife.
● What is it made from? Steel, with plastic handle.
● How was it made? By shaping and sharpening the metal, and moulding the plastic.
● What is its purpose? Cutting and slicing.

This gives a more complete picture of the object, so that it can be fitted into the grand scheme.

The Scientific Kind of Knowledge

In the *Organon*, Aristotle asked: "What is science?" His answer runs: "We have genuine science (*episteme* or true knowledge) when we can state in precise language not only that things are so, but also why they are as they are, and why they have to be that way. We possess science when we can prove and demonstrate statements about them, by relating those statements to other statements of which they are the necessary consequences." It is not enough to believe or know that something is true. We must know how and why.

At each stage in the history of science, people believed that their theories and discoveries were bringing them closer to the truth. We do the same today. It is sensible to suppose that we are not having the "last word". Scientific ideas and theories will continue to change in the future.

Aristotle knew this. He warned many times: "The facts have not been sufficiently ascertained . . . If at any time in the future they are ascertained, then credence must be given to the direct evidence of the senses rather than to theories".

The Final Years

In 323 BC, Alexander the Great of Macedonia died. Aristotle lost an influential friend. People in Athens began to show their resentment for the power of the Macedonians. Aristotle, who came from that area, was one of their targets.

To avoid trouble, Aristotle retired from Athens to Chalcis, on the Aegean island of Euboea (Evvoia). One of the greatest thinkers of all time, he died the next year, 322 BC, apparently of "chronic indigestion".

Aristotle is usually represented as an imposing, thoughtful figure, though we have few details of his personal appearance. Most of his surviving writings are in the form of notes and essays for use while teaching at the Lyceum.

Chapter Six
After Aristotle

The Peripatetic School at the Lyceum in Athens continued after Aristotle. It was led by Theophrastus, who continued the study of metaphysics and psychology, and then Strato, who worked on physical theories. But the school declined as Aristotle's influence faded.

Aristotle left his notes and writings to Theophrastus. The story goes that some were sent to the library at Alexandria, and got lost. Others were hidden in a damp cellar, discovered two hundred years later, and taken to Rome. They were edited, and published in 70 BC by Andronikos, the eleventh head of the Lyceum school.

Aristotle's own writing style was difficult to follow. Since his time the translators, editors and copiers have changed some meanings. In recent times, experts have been trying to separate Aristotle's thoughts from those who have tampered with his works over the centuries.

The library at Alexandria, Egypt, where some of Aristotle's works were kept for a time. The city was founded in 332 BC by Alexander the Great, and had a university and two great libraries.

Aristotle's works have been translated into dozens of languages, with changes and additions happening at each stage during each age. This picture of a money-making mint is from a French version of his works Ethics, Politics *and* Economics.

The Tradition of Aristotle

Aristotle recommended that people should be free to search for the truth, especially by logic and deduction, and also by observing the real world. But during later times, centres of learning altered this approach. They still called it the Aristotelian tradition. But they said that there were given truths, and the role of science was to support and confirm these.

In Europe, ruling groups such as the nobility and the **Roman Catholic Church** developed a stranglehold on scientific progress. Through the Middle Ages, they used their ideas of given truths, of the power of God, to help them stay in power. Not until the time of Galileo and the Renaissance, from about the 1400s, did people begin to question these truths. They began to propose new ideas, and go against the traditional and religious views. Only then, were they able to develop the methods of science and make progress.

The Legacy of Aristotle

During his life, Aristotle dealt with a huge range of scientific subjects, from the physics of motion, to the way grasshoppers chirp by rubbing their legs across their wings. Working within the conditions of the time, he presented a carefully thought-out system to explain the world, and the objects and events in it. The system was so wide-ranging and all-encompassing that it survived for centuries, as philosophers and scientists expanded and revised the contents.

Many of Aristotle's theories and proposals are now out of date. Yet, even as they were being overtaken, his influence remained. Through the ages, Aristotle's deep understanding, his skill at asking the right questions, and his reliance on observation and first-hand evidence, stimulated scientists such as Galileo Galilei, Isaac Newton and Charles Darwin.

Charles Darwin (1809-82, on the left of the picture), the naturalist who wrote about evolution, said that most scientists and philosophers were "mere schoolboys compared to old Aristotle".

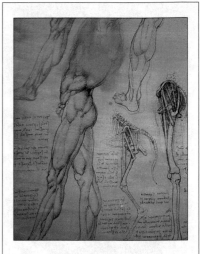

Anatomy

How might Aristotle react if he saw how one of his favourite subjects, anatomy, had progressed? Renaissance artist and scientist Leonardo Da Vinci (1452-1519) produced many anatomical works (above). Today, doctors can use **virtual reality systems** for their basic training in anatomy, without touching a real body (below).

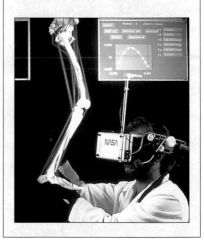

The World in Aristotle's Time

	500-401 BC	400-351 BC
Science	458 Empedocles of Akragos (Sicily) refines the theory of the elements, fire, water, earth and air 450 Leucippus of Miletus, in Turkey, has the idea of the atom, the smallest indivisible unit of matter	400 Hippocrates founds his school of medicine at Cos 400 Lu Pan makes and flies the first known kite, in China 384 Aristotle is born 352 Chinese observers make the first recording of a supernova
Exploration	500 Hecataeus of Miletus produces a map of the world, a flat disc with Europe and Asia surrounded by ocean 470 Hanno of Carthage in Tunisia sails around the west coast of Africa as far as Sierra Leone, and describes the gorilla	380 Eudoxus of Cnidus, in Turkey, works to improve Hecataeus' map of the world 371 Xenophon, soldier and historian, writes *Anabasis* about his travels on campaign with the Greek army in Persia
Politics	510 Roman Republic is founded 480 Greek navy defeats Persian navy at Salamis, ending westward expansion of Persian Empire 431 Athens and Sparta begin the Peloponnesian War	390 Gauls conquer and burn Rome, but it is soon rebuilt 356 Philip II becomes king of Macedonia
Arts	446 Parthenon's foundations are laid in Athens 430 Greek marble sculpture is completed of *Nike tying her sandals* 406 Greek playwright Euripedes dies	387 Plato writes *Symposium* and *Phaedrus* about ideal love 362 Romans begin theatrical productions in the Greek tradition 351 Tomb of Mausolus is completed at Halicarnassus (Turkey), one of the Seven Wonders of the World

Note: *Aristotle lived a long time ago and it is sometimes difficult to say exactly when an event took place. Some of the dates given here are approximate.*

350-301 BC

350 The Chinese create a star-chart system called the *Shan-hai-ching*

322 Aristotle dies

312 Romans build one of their first aqueducts, the Aqua Appia, bringing water to Rome from 16 km away

330 Greek seaman Pytheas sails into the North Atlantic and Baltic Sea, as far as Norway, and around Britain

325 Alexander the Great orders exploration of Indian Ocean, Persian Gulf and Euphrates

312 Work begins on the Appian Way, one of the Roman roads

350 The hillfort, Maiden Castle in Dorset, England, is under construction

336 Alexander the Great becomes king of Macedonia, invades Asia Minor, and conquers Egypt and Jerusalem by 332

326 Alexander the Great reaches India, as far as the River Indus, but dies in 323

350 Theatre at Epidaurus built, of typical Greek design; it is still used today

340 Famed Greek sculptor
330 Praxiteles creates *Hermes with young Dionysus*

330 Alexander's Cenotaph erected at Sidon

300-250 BC

300 Chinese record the use of the lodestone, a magnetic rock, as a compass called a "south-pointer"

287 Archimedes, Greek mathematician, is born

284 The Great Library is founded at Alexandria

300 Dicaearchus of Messana, a student of Aristotle's, produces the first map of the world as a sphere

250 Eratosthenes maps the course of the River Nile, and later estimates the Earth's circumference (he was not far out)

290 Rome completes conquest of central Italy

264 First Punic War between Rome and Carthage in Tunisia begins

300 Work begins on the Colossus Rhodes, a huge statue of Apollo, one of the Seven Wonders of the World

277 One of the first Chinese poets, Ch'u Yuan, dies

264 Gladiator combats become a popular spectator sport in Rome

Glossary

aither: a fifth *element* which Aristotle proposed existed to explain the movement of heavenly bodies. It is also sometimes called ether.

anatomy: the scientific study of the structure of living things, from microbes to trees and from elephants to humans.

atoms: the smallest part of a substance, far too tiny to see under the most powerful microscope. Today atoms can be split into smaller particles, such as electrons and neutrons, but these no longer have the physical and chemical features of the original substance.

deduction: using reason and *logic* to arrive at, or deduce, an idea or fact from a starting set of facts or observations.

dialectic: a branch of *logic;* this is the process of critical reasoning. Studying dialectic involves the examination of the methods and art of discussion and argument.

dialogue: a discussion and exchange of views and opinions, usually between two people or groups.

element: a single, pure substance such as iron or carbon. All the basic parts, or *atoms,* of an element are the same as each other, and different from the atoms of other elements. In the past, people believed there were only four elements, earth, air, fire and water, although some argued that a fifth element, *aither,* also existed.

embryology: the scientific study of how animals develop in the early period of their life, when they are called embryos, larvae or other specialized names. The animals range from a caterpillar growing inside its egg, to a tiny crab larva floating in the sea, to a human baby growing in its mother's womb.

ethics: the area of human thought and behaviour concerned with right and wrong, good and bad, and similar moral questions. It also covers the rules and principles we invent concerning these topics.

evolution: the way in which plants and animals have changed slowly over millions of years from simple to more complex forms of life.

hermaphrodite: an animal or plant that is both male and female, that is, having both male and female sex *organs* in the same body.

hypothesis: a proposal or suggestion put forward to explain facts and events, which can be studied by tests and experiments.

invertebrates: animals without backbones. They range from jellyfishes and worms to crabs, spiders and insects. See also *vertebrates.*

logic: using reasoned thought, analysis and *deduction* to come to decisions and viewpoints, without having to rely on ideas and beliefs that cannot be proven. It is particularly involved with the study of *philosophy* and there are several kinds, such as *syllogism.*

metaphysics: the ideas of existence, knowing and "being", and their substances and qualities. Metaphysics often deals with abstract ideas and notions, in thought only, compared to physics, which deals with objects and events that occur in the real world.

microscope: a device that magnifies very small things, making them look hundreds or thousands of times bigger than they really are.

organ: a distinct part of the body which has a particular function. For example, the brain, the heart and the liver are all organs.

philosophy: the study of human knowledge, beliefs and thoughts. It affects many aspects of our lives, such as how we know things, why we believe in right and wrong, and why we think some things are valuable but others are worthless.

rhetoric: the study of how to use speech, written words and other forms of language effectively, often to persuade others.

Roman Catholic Church: the branch of the Christian Church which has its headquarters at the Vatican, in Rome, Italy, and the Pope as its head.

syllogism: a type of *logic* or *deduction,* of which there are several kinds. They are usually of the type: All As are Bs. All Bs are Cs. So all As are Cs.

vertebrates: in general, animals with backbones. Fishes, amphibians, reptiles, birds and mammals are all vertebrates. See also *invertebrates.*

virtual reality system: a computer-based system which creates a "false world" of sights, sounds and other sensations. These seem real to the user, but they are artificial or unreal ("virtual").

Index